INSIDER TIPS FOR BASS FISHING

JANE KATIRGIS AND
SIMONE PAYMENT

New York

Published in 2020 by The Rosen Publishing Group, Inc.
29 East 21st Street, New York, NY 10010

First Edition

Library of Congress Cataloging-in-Publication Data

Names: Katirgis, Jane, author. | Payment, Simone, author.
Title: Insider tips for bass fishing / Jane Katirgis and Simone Payment.
Description: New York : Rosen Publishing, 2020 | Series: The ultimate guide to fishing | Audience: Grades 5–8. | Includes bibliographical references and index.
Identifiers: LCCN 2019005048| ISBN 9781725347199 (library bound) | ISBN 9781725347182 (pbk.)
Subjects: LCSH: Fishing—Juvenile literature.
Classification: LCC SH445 .K38 2019 | DDC 639.2—dc23
LC record available at https://lccn.loc.gov/2019005048

Manufactured in the United States of America

CONTENTS

INTRODUCTION

On July 9, 1955, David Hayes caught a record-setting 11-pound, 15-ounce (5.4-kilogram) smallmouth bass in Dale Hollow Lake, Tennessee. His record still stands today. The closest catch since then was a full pound less than Hayes's trophy bass.

Not all anglers are fishing for a record-breaking catch, though. For most people, fishing is an easy and relaxing pastime. The time spent outdoors—enjoying fresh air, nature, and wildlife—can be just as enjoyable as the sport of fishing itself. Fishing can be enjoyed with others, or it can be a quiet time to spend on your own. Because bass are found in lakes, ponds, streams, and rivers, it is possible to fish in many different bodies of water all around the United States and Canada. Each fishing outing can prove to be unique, too. The unpredictable nature of the hobby is an added benefit. Anglers can never be sure what they are going to catch!

There are several species, or kinds, of bass; some are freshwater species and others are saltwater fish. Striped bass live in ocean waters along the coasts. Largemouth and smallmouth bass can be found in lakes, ponds, rivers, and streams in most parts of the United States and Canada. Other types of bass—such as spotted bass, white bass, and white perch—are also found throughout the United States and Canada.

The habits and diets of the different bass dictate the type of bait or lures anglers will use when fishing. Anglers use simple equipment, including fishing poles, fishing line, and hooks to catch bass. The gear is not too expensive and, if cared for properly, lasts for many years.

Fishing is a relaxing hobby that allows anglers to spend time with friends or family in nature.

As with any sport, fishing has rules and regulations. It is important to understand the water conditions and to be safe near the water or on a boat. Make sure to take care when handling sharp hooks. Good sportsmanship is expected from all anglers, too, as is respect for the animals and the environment. If you practice catch-and-release fishing, make sure to give the fish the best chance for survival.

CHAPTER ONE

GETTING STARTED

People who enjoy fishing as a hobby often study the fish's habitat, life cycle, and behaviors. For example, it is difficult to fish for a species without knowing where it lives or when it eats. Is it a freshwater or saltwater species? Does it feed at night or during the day? Finding out about a fish's diet also helps determine the bait to use. Learning about fish that live in nearby waters can give anglers ideas about what species to try to catch.

GILLS, SCALES, AND FINS

Fish are cold-blooded, vertebrate animals. Vertebrates have a backbone (vertebra), and their body temperature adjusts to match the temperature of their surroundings. Fish breathe through gills, which are flaps of skin located on either side of the head. Fish must move through the water for the gills to work. As water flows through the gills, they absorb oxygen from the water.

A fish's fins and muscles power it through water. Scales protect the body, and a coating of mucus over the scales offers defense against infection. The mucus coating also helps fish move through water quickly.

To help them find food, fish have well-developed senses. They have excellent senses of smell and hearing. They are able to taste with their mouth and tongue but also with the exterior of their body. Fish can also sense food via vibrations in the water. Fish are able to see well. Most can see in many directions because their eyes are near the top of their heads.

MORE THAN ONE TYPE OF BASS

There are several types of bass. They are so different that some bass live in fresh water and others live in salt water. The two major types of freshwater bass are largemouth and smallmouth. Other bass that live in fresh water are spotted bass, white bass, and white perch. Striped bass begin their lives in fresh water, but as adults they spend most of their time in salt water.

LARGEMOUTH BASS

Largemouth bass are the biggest type of bass. They are a very popular North American fish, partly because they put up a good fight, which can make them a challenge to catch. Light green or brown on top, largemouth bass have a white stomach with dark spots shaped like diamonds on their sides. Largemouth bass

Largemouth bass

Smallmouth bass

are usually between 5 and 10 pounds (2.3 and 4.5 kg) when full grown, but can weigh as much as 20 pounds (9 kg).

SMALLMOUTH BASS

Smallmouth bass are the second largest type of bass. Like the largemouth, they put up a good fight while being caught. They are usually a bit thinner and smaller than largemouth, and they have a smaller mouth. Smallmouth bass are brown, gold, or olive green. They are lighter on their sides and have a white belly. Unlike the largemouth, smallmouth bass have red eyes. Smallmouth bass range in size, but can be from 2 pounds to 6 pounds (1 to 3 kg) when full grown.

MORE FRESHWATER BASS

There are several other types of freshwater bass. The spotted bass is often confused with the largemouth bass because they look similar. However, spotted bass are smaller. Other ways to tell them apart are that spotted bass have teeth on their tongue and a large spot near their gills.

Spotted bass

White bass also put up a good fight and are a good fish for eating. They are usually silver with a dark gray or green back. They have yellow eyes and several dark stripes that run along their sides.

Although the name suggests they are part of the perch family of fish, white perch are actually bass. Like white bass, they are very good for eating. White perch look similar to white bass but have a thin body. They can be olive, grayish green, brown, or even black on their backs, with light green or white on their belly. Although most white perch live in fresh water, they can also live in salt water or brackish water, which is a mixture of salt water and fresh water.

White bass

White perch

STRIPED BASS

Striped bass are large and have a long body. They are bluish black or dark green on top with a silver or white belly. Black stripes run along their sides.

Unlike most types of bass, striped bass live most of their lives in salt water. As adults, they live in coastal ocean waters. But when female striped bass are ready to

Striped bass

FRESHWATER STRIPED BASS

Some striped bass live their whole lives in fresh water. Some of these freshwater striped bass were moved by people to inland freshwater lakes and rivers, and survived to breed and start new populations. Other freshwater striped bass populations began when the Santee and Cooper Rivers in South Carolina were dammed in the 1940s. Striped bass in those rivers at the time were trapped but adapted to the conditions.

lay eggs, they swim up freshwater rivers connected to the ocean. After they lay their eggs, they swim back out to the ocean. Any eggs that aren't eaten by other fish or predators hatch in the river. The hatchlings spend their early lives in the river. Eventually, the young striped bass swim out to the ocean.

WHEN TO BASS FISH

The best time to fish varies by location. Asking local anglers when they have had luck can be helpful. Many fishing websites can also provide useful information. Some state and local environmental agencies post local fishing information on their websites.

In general, the cooler weather of spring and fall is best for freshwater bass, but bass can also be caught in the summer. In some southern locations, bass can be caught in the winter months, too.

Cloudy or overcast days usually make for ideal fishing weather. Another good time is just before it rains. However, it is not usually good to fish after rain or when it is windy because the water is stirred up too much. Most bass prefer to feed when the water is clearer. Early to mid-morning is usually the best time of day to fish. The last few hours before sunset can also be a productive time to fish.

FISHING FOR LARGEMOUTH BASS

Largemouth bass live in many types of bodies of water throughout the mainland United States and a few southern areas of Canada. They can be found in small creeks or ponds, but more often they live in lakes, larger ponds, and rivers. They like fairly warm water, of

Sunrise and sunset are ideal times to go fishing.

about 70 degrees Fahrenheit (21 degrees Celsius) or higher.

Largemouth bass occasionally look for food near the surface of the water. However, largemouth bass usually stay well below the surface of the water. They especially like darker water, so they are often found in shady areas, such as below branches or trees that have fallen into the water. They can also be found near low-hanging branches or underneath lily pads. However, largemouth bass will usually be found only in these areas if the water is a few feet deep or more.

FISHING FOR SMALLMOUTH BASS

Smallmouth bass prefer clearer water than largemouth bass. They also like water with a rocky bottom, so they are most often found in streams or clear lakes. They are also happier in cooler water than largemouth bass—about 65°F (18°C). Smallmouth bass can be found throughout most of the United States. They are also widely found in Canada.

Smallmouth bass inhabit clear waters with rocky bottoms.

FISHING FOR OTHER FRESHWATER BASS

Spotted bass live most often in streams. White bass live in large lakes or rivers. White perch can be found either in salt water near the shore and coastal rivers or in freshwater lakes and ponds. White perch like to be close to the bottom of any body of water.

FISHING FOR STRIPED BASS

When they are in the ocean, striped bass are usually near the shore. They like to look for food around bridges, docks, piers, and rocks. Striped bass spend most of their time looking for food around these structures and near the bottom, but they will come to the ocean surface if they are chasing food. When striped bass are in fresh water to lay eggs, they are usually in the open water in the middle of a river.

The largest populations of striped bass are along the East Coast of the United States. However, there are smaller populations along the Gulf Coast and in the Pacific Ocean off the coast of the northern United States and southern edge of Canada.

BAIT FOR BASS

Two types of bait can be used when fishing for bass: live bait and lures. Small fish, insects, or other things that a fish would eat in its natural environment are examples of live bait. Live bait can be purchased from a bait or fishing shop or can be caught. Lures are artificial but are made to look somewhat similar to natural bait. Bait shops and local anglers can offer good advice on what kind of live bait or lures to use.

USING LIVE BAIT

Live bait includes worms, crickets, grasshoppers, crayfish, and minnows. It can be found at bait shops and sometimes at fishing shops or other local stores. In some areas, bait can even be purchased from vending machines.

Bait can also be caught easily in most areas. Worms can be found in nearly any yard, especially after a rain or at night. To keep worms alive and healthy until it is time to fish, make sure they are cool and damp by adding some soil, moss, or damp newspaper to the container, and leave air holes in the container.

Crickets or grasshoppers can be found in tall grass and caught with a small net. Keep insects cool and dry with some grass or dry newspaper in a container with plenty of air holes. Crayfish can be caught in shallow water near rocks with a net. Minnows can also be caught in shallow water with a small net or minnow trap. Store crayfish or minnows in water until ready to use.

When using live bait, put it on the fishing hook in a way that won't kill the bait. This way, it will still move in the water and attract a fish. At the first hint of a nibble on the line when using live bait, tighten the line to "set the hook" in the fish's mouth. Waiting too long will allow the fish to possibly swallow the bait.

USING LURES

Lures can be made of several types of material. Some look like minnows and are made of metal. Some look like worms, frogs, or other animals. There is also a type of artificial bait called flies. These are made of thread and other fiber, and they mimic insects. Lures and flies can be purchased at local bait or fishing stores or online. When using lures and flies, set the hook quickly, at the first bite from the fish. Otherwise, the fish may realize the lure is artificial and will lose interest.

Artificial bait that looks like insects or small fish move when pulled through the water and make a fish curious enough to attack them.

BAIT AND LURES FOR FRESHWATER BASS

Largemouth bass eat a wide variety of food, so many types of bait can be used. They most often eat smaller fish and crayfish. They will also eat frogs, mice, salamanders, snakes, and worms. Live bait like minnows and crayfish will attract largemouth bass, but lures will usually work as well.

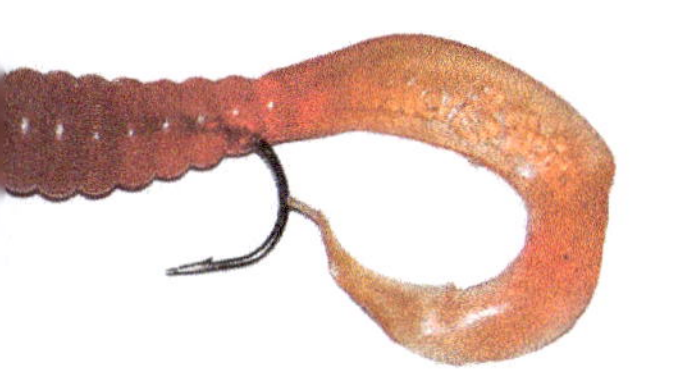

Smallmouth bass eat whatever they can find. Their favorites are small fish and crayfish, but they will also eat worms, ants, grasshoppers, ladybugs, leeches, and snakes. Like largemouth bass, smallmouth bass can be caught with live bait or lures.

Spotted bass eat insects, minnows, frogs, worms, and small fish. Crayfish, however, are their favorite food source. White bass eat smaller fish and a variety of insects and crayfish. White perch eat small fish, crabs, and shrimp.

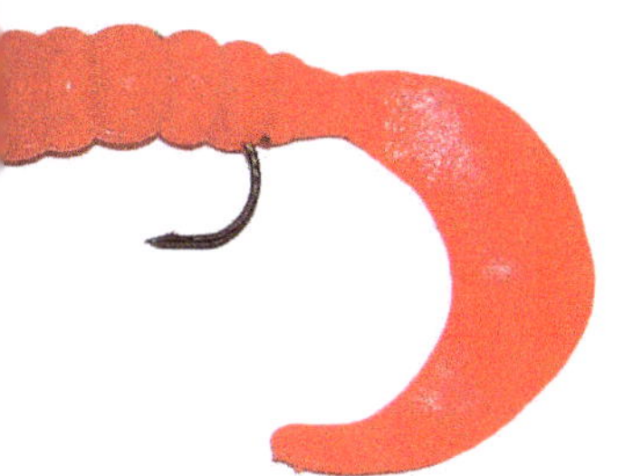

BAIT FOR STRIPED BASS

Striped bass are not picky eaters. They will eat smaller fish, squid, worms, and crabs. They are happy to eat live bait but will eat dead bait, too. Although eels and clams aren't what they usually eat, chopped eels and clams can be used to attract striped bass.

CHAPTER TWO

GEARING UP FOR BASS FISHING

One of the pleasures of a fishing hobby is that there is not much equipment needed. A fishing pole, fishing line, and bait or lures are the only necessary requirements. Of course, as with any hobby, there are extras that could be purchased, but these are usually unnecessary. There are a few things any angler will need, though.

FISHING RODS AND OTHER EQUIPMENT

A fishing pole is a piece of wood with fishing line and hook attached to the end of the pole. A pole can be as simple as a tree branch or piece of bamboo. A pole is easy to make, or it can be purchased. Because there is no reel to control the fishing line, with a pole an angler flicks the line into the water. When a fish bites, simply pull the fish out of the water.

Rods are human-made, and they have a spot to attach a reel of fishing line. Guides run the length of the rod to keep the line in place. Rods are usually made of fiberglass.

Fishing line is usually made of nylon and is a single strand (called a monofilament). Most fishing line is clear, but it comes in

A spin-casting fishing reel holds the fishing line and sits on top of the fishing rod.

other colors including red, blue, green, and yellow. There are different strengths of fishing line, and each strength can hold up to a certain amount of weight. The number on the package tells how many pounds a particular line can hold.

There are several types of reels to hold fishing line. A spin-casting reel is the easiest to use. It attaches to a fiberglass rod, and the fishing line is enclosed within the reel. Spin-casting reels sit on top of the fishing rod.

Spinning reels take a little more practice to use. The fishing line is not enclosed in a spinning reel, so it can get tangled. The advantage is that it is possible to cast farther out into the water with spinning reels. They attach to fiberglass or graphite rods and sit on the underside of the rod.

Professional bass anglers often use bait-casting reels. The reel winds the line from side to side, not around in a circle. The advantage of using a bait-casting reel is an angler can be more specific about when to release the line. However, the fishing line can easily get tangled, making this type of reel difficult to use. Bait-casting reels are used on fiberglass or graphite rods and attach to the top of the rod.

There are two types of hooks: single and treble. Single hooks, which are used for bait, have just one hook. Treble hooks have three hooks and are most often used with lures.

Hooks are sold with and without barbs, which are sharp points at the tip of the hook. It is best to buy barbless hooks because they cause much less damage to the fish. If barbless hooks are not available, file the barbs off of the hook with a metal file. Or bend the barb "closed" using pliers.

MORE GEAR

In addition to the gear an angler will absolutely need to have, there is other equipment that can make fishing easier. Snap swivels are clips that are attached to the end of the fishing line. These swivels allow anglers to easily put hooks or lures on the line without having to cut the line or tie new knots each time. The snap swivels move around freely so that the fishing line does not become twisted when hooks and lures move through the water.

Sinkers attach to the fishing line above the hook to help the hook sink to the bottom. Bobbers have the opposite effect. They can keep part of the line afloat so the hook does not sink too deeply. Bobbers can also alert the angler when

LOCATION, LOCATION, LOCATION

When deciding where to fish, the first thing to do is check local fishing regulations. Fishing is prohibited in some areas and bodies of water. There can also be restrictions based on the season, so a fishing area may be open at certain times, but not others.

A great way to locate a good fishing spot is to ask a local angler. He or she will be able to point out not only a good body of water, but possibly even the best spot on that body of water. To find a local angler, talk to friends or family, or ask people at a bait shop or store that sells fishing gear.

If a fishing spot turns out to be a good one, make a note of its location. Also note what time the fish were biting, the weather conditions, and the bait used. This information can help when planning a future fishing trip.

a fish hits the hook because they move on the surface of the water.

Tackle boxes store the small equipment—hooks, lures, sinkers, snap swivels, bobbers, and so on—an angler needs on fishing expeditions. Tackle boxes can be especially handy when using lures. They keep lures separated and can prevent hooks from getting tangled. Tackle boxes are available in stores, but it's also possible to make one out of a box with an egg carton inside it. Small squares of Styrofoam make good storage material for hooks.

A knife for cutting bait or line is useful. Gloves are also usually something an angler will need. Cotton gloves that can be dipped in

An organized tackle box can hold an angler's tools, hooks, lures, and fishing line. It is helpful to have all the items in one place.

water are good for handling fish when removing hooks. Thicker gloves can be worn when cleaning fish in preparation for eating.

Some anglers use nets to lift fish out of the water. However, nets can cause a lot of damage to fish. If a net must be used, plastic nets are the best type. These cause much less damage to the fish's protective mucous coating.

SONAR, GPS, AND BOATS

Sonar and the Global Positioning System (GPS) are two types of technology that can be very helpful when fishing. Sonar is also called a depth finder, fish finder, or depth sounder. It shows the underwater environment on a video screen, allowing anglers to "see" the bottom, large objects, and schools of fish. GPS is a satellite-based navigation and location system. Users receive signals on their devices that can help them determine their exact location.

Fishing from a flat-bottomed boat allows anglers to navigate shallow waters in search of bass.

GPS systems used for fishing are the same devices that are used in a car. A GPS can help find a specific fishing area if coordinates are already known. Another good use of GPS is "marking" a productive fishing spot. Recording the fishing spot's coordinates will make it easy to find on the next fishing expedition. Some GPS systems are so small and portable that they can be used even on rowboats or canoes.

Bass fishing can easily be done from the shore, a dock, or a riverbank. But many anglers fish from canoes, rowboats, or small engine-powered boats. There are also boats specifically designed for bass fishing. These are most often used by professional bass anglers or other serious anglers. Bass fishing boats are usually very fast, fairly flat on the bottom, and made of fiberglass. The flat bottoms allow the boat to go faster and get close to shore in shallow areas. Bass boats have low, flat deck areas in the front and back to make it easy to fish on all sides of the boat. Flat decks also allow anglers to fish standing up. Bass boats usually have a lot of storage space below the deck and are equipped with sonar and GPS.

STAYING COMFORTABLE AND SAFE

Waders can be worn when an angler will be standing in a stream or a lake to fish. They keep an angler dry and can offer some protection from cold. They can also make it easier to walk on the slippery or muddy bottom.

Taking rain gear or a warm jacket on a fishing expedition is always a good idea, even when rain is not expected. Rain can occasionally arrive unexpectedly, and weather conditions can change quickly on the open water of a large lake or the ocean.

One of the most important pieces of safety equipment any angler should have is a personal flotation device (PFD), or life preserver. Anglers should always wear one in a boat or while wading in a stream or river. Even good swimmers should wear a PFD.

CHAPTER THREE

BASS FISHING SPORTSMANSHIP: RULES AND SAFETY

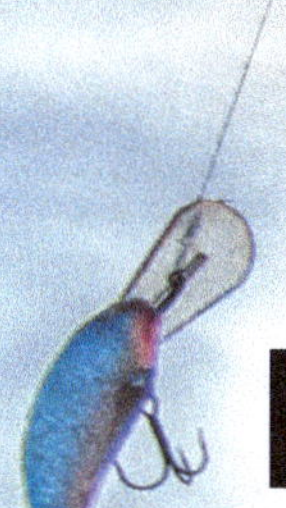

Although fishing should be enjoyable, relaxing, and exciting, responsible angling is one of the most important aspects of the sport. The responsibilities include personal safety, respect for wildlife and the environment, and good sportsmanship. Poor sportsmanship and irresponsibility can result in damage to the environment, declines in fish populations, and, eventually, fewer opportunities to fish.

THINK BEFORE YOU ACT

Using common sense should be first and foremost. Be aware of local rules and be aware of the surroundings. This awareness includes staying off private property—land, docks, and beaches. Don't get too close to other anglers when fishing from shore. In a boat, don't get too close to docks, the shore, wading anglers, or other boats.

Another aspect of sportsmanship is taking care of the natural environment. At the end of a day of fishing, it is essential to bring all hooks, fishing line, bait containers, and other garbage home to dispose of or recycle it properly. Collecting hooks and fishing line is especially important. Fish, shore birds, and other aquatic

Loose fishing line is a serious hazard for wildlife. Make sure to collect any line you cut and take it with you to throw away.

creatures like turtles or salamanders can swallow hooks or become tangled in fishing line. This situation can kill them or prevent them from finding or eating food.

Another way to protect the environment is to clean fish only at home or in specific fish-cleaning stations provided on some docks and marinas. Don't clean fish on a public dock or even along the shore.

Practicing good sportsmanship also means being ethical. Ethics is not the same as legal versus illegal. Instead, ethics concerns what is right versus what is wrong. Just because there is no specific law against something doesn't necessarily mean it is right to do it. For example, most anglers consider it unethical to catch and keep fish that are laying eggs. This is because it interferes with egg laying and eventually lowers the population of the fish. In most states, it is not illegal to catch fish while they are laying eggs. But in some locations, waters where fish are known to lay eggs are closed to fishing during the spring.

Catching too many fish is also considered unethical, even if a particular angler has kept only the limit allowed. If every angler keeps the limit on every fishing trip, it could be harmful to the long-term survival of the fish population.

Another way to practice ethical angling is not to pollute fishing waters or any other area. Even if there is no law in place to prevent a specific type of pollution, it is in everyone's best interest to keep bodies of water and the surrounding environment clean.

BE SAFE ON THE WATER

Because fishing takes place in and around all kinds of bodies of water, good water safety practices are essential. Perhaps the most important rule of water safety is: always wear a PFD. It is not enough to have a PFD on the boat or nearby. It can be too difficult and sometimes too late

Always wear PFDs when fishing from a boat. Even good swimmers need to practice safety at all times.

to put one on if a boat capsizes. Even good swimmers should always wear a PFD. When buying a PFD, make sure it fits snugly but not too tightly.

Knowing how to swim is a valuable skill for any angler. Many schools or recreation programs offer swimming lessons. If possible, take water safety classes at school or from a local Red Cross or other program. Water safety classes can teach procedures for how to help someone who has fallen overboard or how to flip an overturned canoe.

KNOW YOUR BODY OF WATER

Anglers should also be familiar with any body of water being fished. This familiarity is especially important in rivers. Be aware of the current and how strong it is. Sometimes, the surface of a river or stream looks calm, but underneath the water is moving quickly. Watch out for objects like logs or branches floating downstream.

Also be alert when fishing on the shore of the ocean. Sometimes, particularly near rivers that flow into the ocean, the tide can come in or out quickly. When this happens, water can rise or lower swiftly, leaving anglers in unsafe conditions.

The bottom of most bodies of water is often slippery, so take care when walking. Take small steps, planting feet firmly. Try to put feet between rocks rather than on top of rocks because rocks can be covered with algae or other slippery material.

In a lake, be aware of how deep the water is before wading in along a shore. Use a stick to find the bottom of the lake to measure the depth.

BOAT SAFETY

Being careful in and around boats is another aspect of water safety. Don't overload a boat with people or equipment. Make sure there is safety equipment on board. This equipment should include a first-aid kit, lights, a horn, and PFDs for everyone on the boat. When navigating, be alert for other boats, obstacles, dams, rocks, and sandbars. Knowing an area well before taking a boat trip is a good idea. Also watch the weather conditions. Strong winds, high waves, and thunderstorms can be dangerous for any size boat.

WEATHER WATCH

Whether in a boat or on shore, watching out for thunderstorms is extremely important, especially during the summer months. Check the forecast before heading out for a day of angling. If thunderstorms are in the forecast, consider changing the day or time of the trip. Thunderstorms can also come up quickly, so even when a thunderstorm is not in the forecast, keep an eye on the skies. Using radios or electronic devices like smartphones to get updates on the forecast while angling is a good idea.

If thunder can be heard, don't delay: take shelter. A building or a car is the best place for shelter. If neither is available, try to find the lowest place and stay near the smallest trees. If in a field,

BOLTS OF LIGHTNING

A warning sign that a lightning strike is imminent is that the hair on a person's arms or head will stand on end. If this happens, kneeling with hands on knees, bent forward is the best position. This position gives lightning the least amount of contact with the ground.

stand in an open area but away from tall trees, telephone poles, or metal fences or posts. Take off anything made of metal.

FISHING GEAR SAFETY

Some fishing equipment is sharp, so be safe and alert when working with hooks and knives, for example. Even parts of fish, such as fins, gills, and teeth, can be sharp, so take precautions. Only use a knife with clean, dry hands. Or wear gloves for extra protection. Be alert when using a knife, especially in a boat, which can move suddenly. Also be alert when someone else is using a knife to avoid coming in contact with the knife by accident.

Using barbless hooks is a good idea not only for the sake of the fish, but also for the angler. Barbless hooks are less sharp. Barbless or not, always secure a hook on the line guide of a pole and make sure the line is tight when the pole is not in use—or remove the hook from the line.

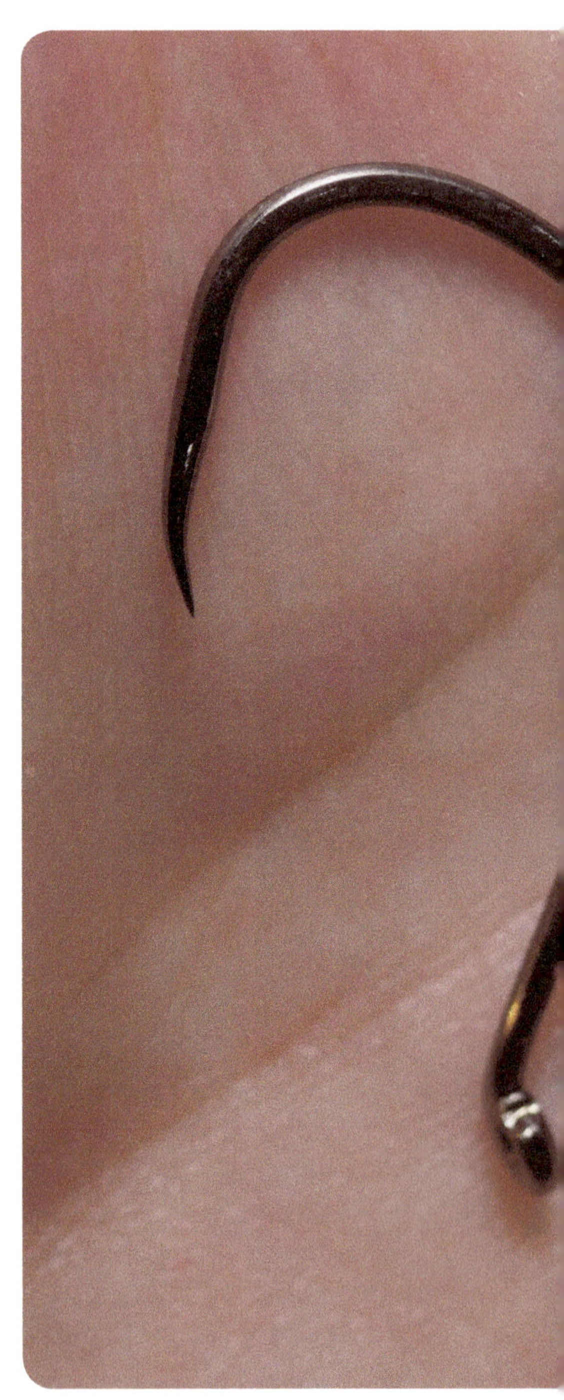

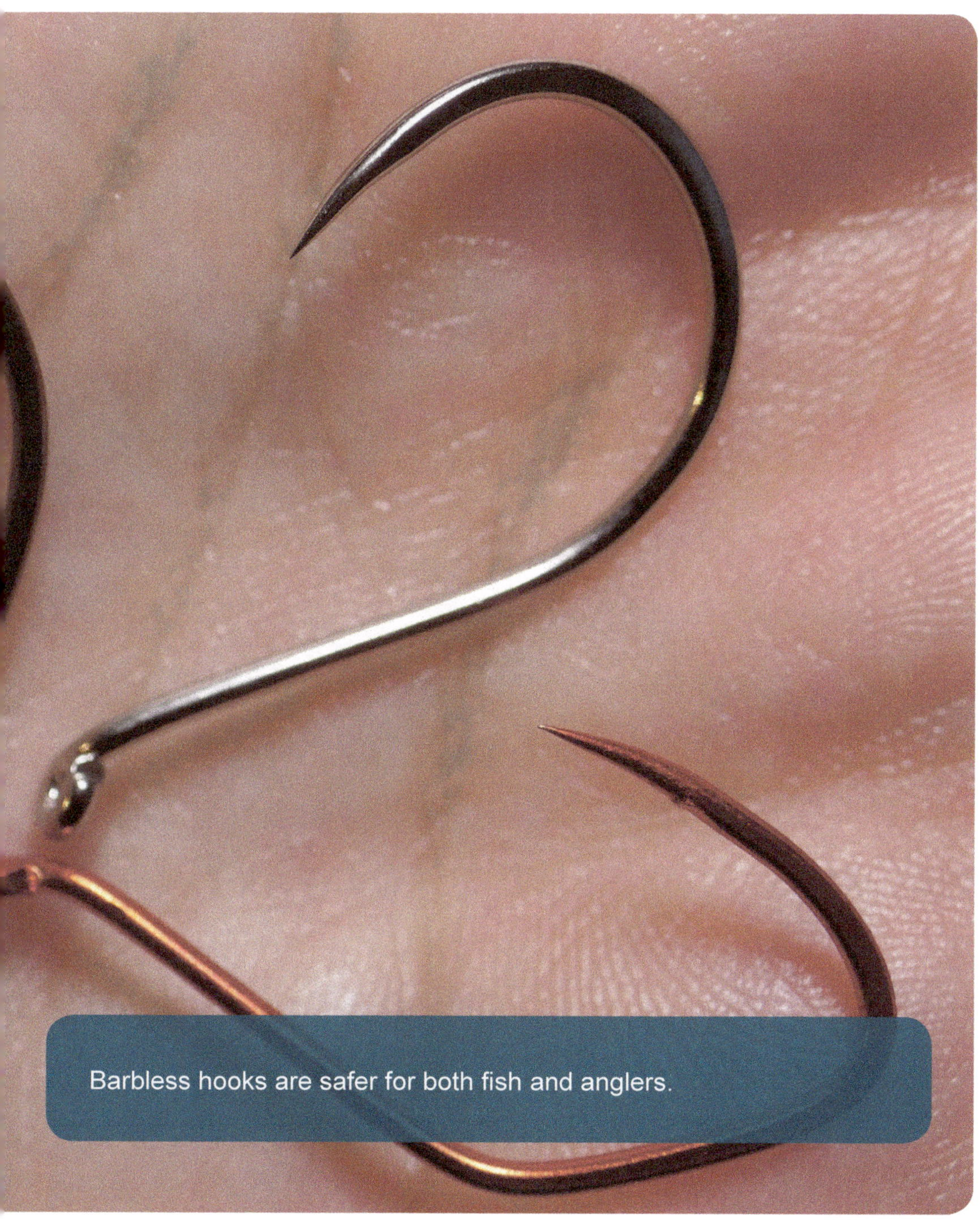

Barbless hooks are safer for both fish and anglers.

When handling fish, avoid touching fins, which can be sharp. Gills can also be sharp. Wearing cotton gloves when handling live fish, as mentioned previously, protects the fish. But even dead fish should be handled with gloves to avoid cuts and scrapes.

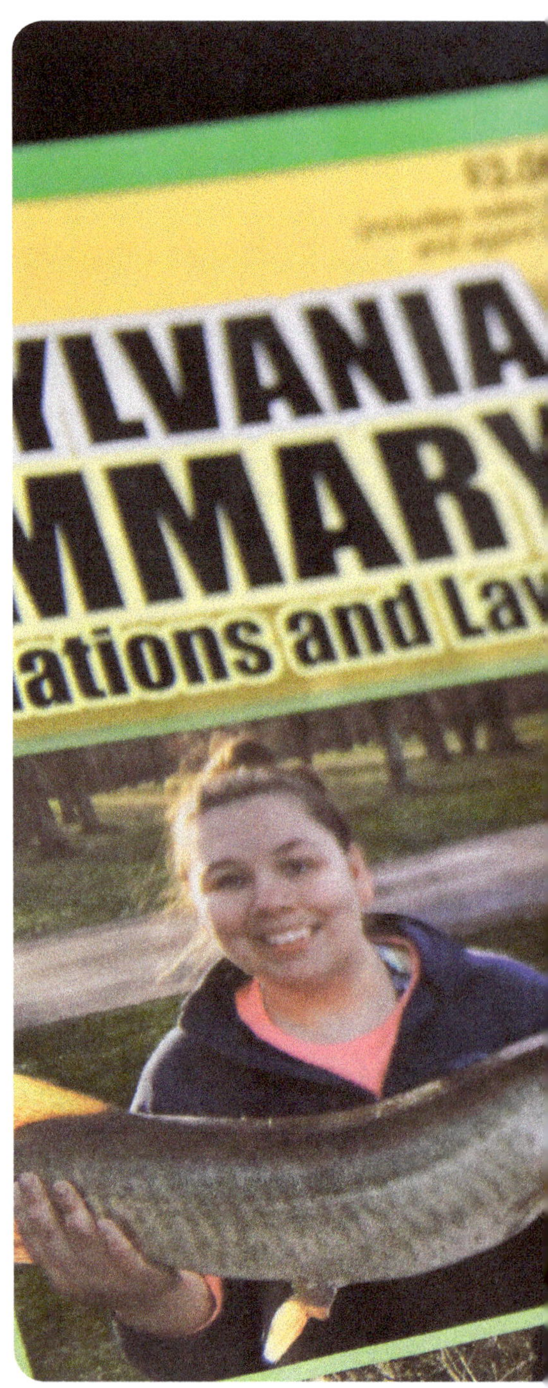

LOCAL LAWS AND REGULATIONS

Every angler should be familiar with local laws and regulations. It is the responsibility of the angler to know which fish it is permissible to catch and if there are size or amount limits on those fish. Anglers must also know whether or not a license is required for fishing.

In some states, it is illegal to catch a particular type of fish. If caught, those fish must be released immediately and in good condition. This law is usually to protect a rare or endangered type of fish. Limits might also be set to protect fish with a low population in a particular area.

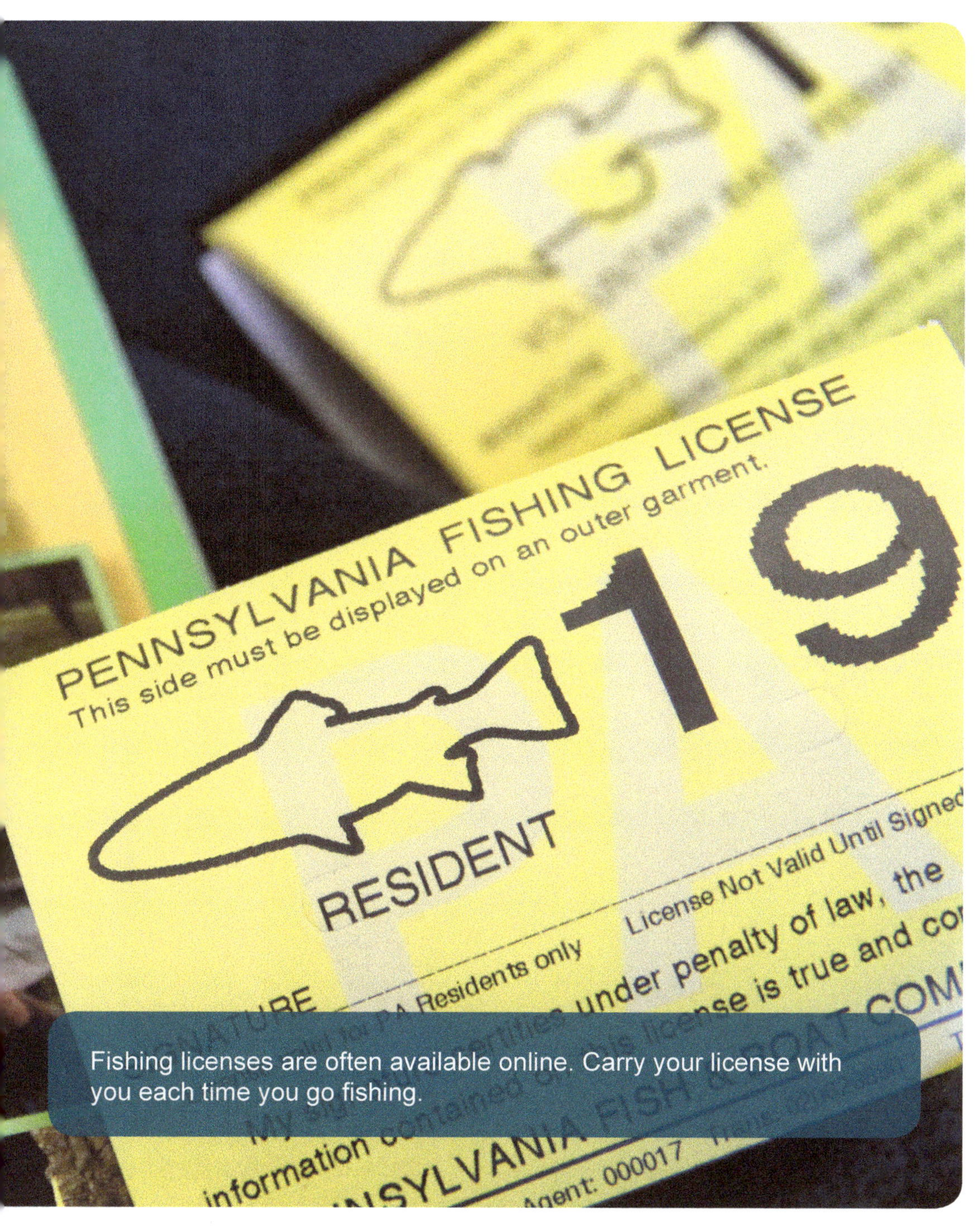

Fishing licenses are often available online. Carry your license with you each time you go fishing.

Some states set length limits. For example, a law might require that any fish under 6 inches (15 cm) in length be released.

Many states put limits on the number of fish of a certain type caught in one day. Other laws might apply to when fish can be caught. Some fish have seasons when they cannot be caught. This situation does not apply to most types of bass, except striped bass in some cases.

Fishing licenses help fund conservation and other state efforts to protect fish and enforce rules. Each state has its own fishing license, and anglers must be licensed in the state where they want to fish. Some states issue different permits for saltwater and freshwater angling. Anglers under a certain age may not need a license. There are usually reduced license prices for children and teens if a state does require a license. Licenses are good only for a certain length of time. Some states issue licenses that are valid for one season, or for one full year. Others issue licenses for one calendar year.

Fishing licenses can be purchased from local department of fisheries and wildlife offices. Most states also allow anglers to buy their licenses online.

CHAPTER FOUR

AFTER THE CATCH

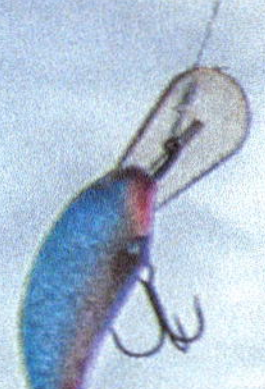

Catch-and-release fishing is the practice of returning fish to the water after they are caught. This practice allows them to live and reproduce, resulting in a strong fish population. Not all fish survive when released, but giving the fish a chance to live is better than not. Catch-and-release practices done properly can give a fish a better chance of survival. If an angler decides to keep the fish, it should be stored, cleaned, and cooked properly.

THE RULES OF CATCH AND RELEASE

There are both official and unofficial rules about keeping a fish versus throwing it back. Some anglers have their own unofficial rules about when to keep or not keep a fish. For instance, some never keep fish. Some keep only a certain number per month or year. Others make their personal rules based on the size or rarity of a fish. Many anglers always return rare or particularly big fish to the water. These larger, older fish are better able to breed and increase the fish population. Some anglers make the decision to catch and release a big fish out of respect for the fish and its ability to survive and reach a large, healthy size.

Official catch-and-release policies vary by type of fish or even by the body of water. For example, anglers might be allowed to keep a fish of a certain length from one lake, but not a fish of the same length from a lake just a mile or two away. Environmental officials usually make policies based on the fish population in a particular location.

SURVIVAL AFTER CATCH AND RELEASE

Most fish can survive when they are released back into the water. However, just the stress of being caught or handled can cause fish to die. There are some factors, or a combination of factors, that can make it more likely that fish will die. Avoiding the following things can increase a fish's chances of survival. First, because fish breathe through their gills, an injury to a fish's gills, even a minor one, is something anglers should try to avoid. Keeping the fish out of the water for as short a time as possible is also essential.

A long fight can cause a substance called lactic acid to build up in a fish. This effect makes it difficult for the fish to process oxygen. So, bringing the fish to the surface with a minimum of fighting time is a good practice.

The kind of bait used can also affect a fish's chances of survival. Live bait is usually worse for fish because they swallow it more deeply and the hook then causes more internal damage.

Although the angler is not able to control this, where a fish is hooked also makes a difference in its survival. Fish hooked in the mouth, jaw, or cheek generally have the best chances. Fish hooked in the eye or gills usually have a decreased chance of survival.

When practicing catch and release, anglers should minimize the time they spend fighting a fish after it is caught.

HOW TO CATCH AND RELEASE

There are several things an angler can do to increase a fish's chance of survival after being released. Most important, if at all possible, keep fish in the water when removing the hook. A good rule of thumb is that fish should not be out of the water longer than twenty seconds at a time. This guide is probably easiest to follow when fishing from shore, rather than on a boat. But

Handle a fish gently, and hold it by its tail or mouth.

even on a boat it is possible to leave the fish in the water when removing a hook.

Most bass are strong and put up a good fight when being caught. But it is usually easy to handle them because they don't squirm too much. Keep the fish still by gently holding its tail or, preferably, its jaw. (Remember to avoid touching a fish's gills.) If possible, anglers can work in teams with one angler holding the fish and the other removing the hook.

Most of the time, hooks usually do not go in too deeply in bass, so they are fairly easy to remove, especially if they're barbless hooks. To remove the hook, use a pair of long-nosed or needle-nosed pliers to gently ease the hook out of the fish's mouth. It is sometimes possible to simply put some slack in the fishing line for the hook to come right out. If the hook does not come out right away using pliers or slack in the line, don't tug or pull hard. This action can rip the fish's mouth or gills.

If a fish is deeply hooked, some experts recommend cutting the line and leaving the hook inside the fish. This works best if the hook is in the throat and not all the way into the stomach. However, leaving a hook inside a fish will usually greatly reduce the fish's chances of survival. This case is especially true for saltwater fish like striped bass because the hook can rust in the salt water.

Although working quickly is important, it's also important to work carefully when removing the hook. It is easy for a hook still attached to a wriggling fish to become caught on a piece of clothing or a hand or finger.

It is not always possible to keep the fish in the water when removing the hook. If this is the case, pick the fish up gently while wearing damp cotton gloves. Or, put a damp towel or cloth over the fish's head or body to handle it. Wearing damp gloves or using a towel does less damage to the mucus coating that protects the fish from infection. Keep a gentle, but firm hold on the fish. Don't let the fish flop around on the ground or the bottom of the boat. This occurrence can cause a lot of damage to the fish. It also makes it more likely that the hook can become caught on a person.

Once the hook is removed, the fish can be released. If the fish is already in the water, simply let go of the fish and it will swim

Gently release a fish back into the water. If the fish needs reviving, move the fish forward in the water to get water to flow across its gills.

away. If the fish is not already in the water, never throw or toss the fish back in the water. Instead, gently place it in the water.

If the fish is not moving much, it may be stressed. If this is the case, hold it gently in the water, keeping it upright. Moving the fish forward will allow water to flow through its gills. This step can help revive it. If the fish is in a river, let the fish go with the current, not against it. When the fish begins to move on its own, gently release it.

CHOOSING TO KEEP THE FISH

It is best to practice catch-and-release fishing. It is more humane and better for the health of the fish population. However, in

COOKING AND CONTAMINATION CONSIDERATIONS

When deciding whether or not to eat a fish, keep in mind that it is best not to eat too many fish from the same area. This stipulation is because some waters contain contaminants that are stored in a fish's body. These contaminants, especially mercury, can be harmful to humans. Check with state or local environmental offices about local bodies of water and safe amounts of fish to eat from them.

Bass are not too fatty, but it is still a good idea to get rid of as much fat as possible because the fatty parts of fish are where contaminants are stored. Good cooking options for burning off fat are grilling or broiling. Special grilling baskets or pans can make grilling a fish easier. Fresh fish tastes so good that it's usually only necessary to cook it with a little olive oil or butter and some salt and pepper.

some cases, an angler might choose to keep a fish to eat, as long as it is permitted by local fishing regulations. Be sure to check local sources to make sure it is safe to eat fish caught in a particular body of water.

HOW TO STORE THE FISH

To prepare a fish for eating, the first priority is keeping the fish cool until it can be cleaned and cooked. Put the fish on ice as soon as possible after it is caught. The best way to store the fish is in a cooler with ice and a bit of water. This step will keep the fish not only cool but moist as well.

HOW TO PREPARE THE FISH

Fish should be cleaned as soon as possible. Clean fish at home or in a designated fish-cleaning station at a dock or marina. These stations will have water and facilities to dispose of guts and skin and bones. These fish parts smell and attract flies and other insects, so they must be stored and disposed of quickly. Never clean a fish on a beach or dock or other public area.

When cleaning fish at home, make sure to dispose of the discarded parts quickly. Wrap them well in plastic bags that can be sealed or knotted, and put them in tight, outdoor garbage cans. The smell of dead fish attracts animals, such as raccoons, bears, dogs, or coyotes, so use garbage cans with tight lids.

Cutting a fish for eating is called filleting. Filleting separates the skin and bones of a fish from its flesh. It is not an easy thing to do, but with some practice, it gets easier. However, it is best to have help from an adult or other experienced person for the first few attempts at filleting.

A sharp knife is essential for filleting. Wear a pair of gloves for protection from sharp parts of the fish and from the knife.

Many anglers decide to fillet their fish and enjoy a fresh and healthy meal.

Also, fish tend to be slippery, so wearing gloves can provide a firmer grip.

When filleting, start where the head meets the body. Begin cutting along the back, about 1 inch (2.5 cm) deep to reach the backbone. Pull the meat away from the ribs toward the stomach. Then cut the skin of the stomach down toward the tail, leaving the tail on. Turn the fish over and hold the tail. Slip the knife between the skin and the meat of the fish, and work from tail to head. Repeat this process on the other side of the fish.

CHAPTER FIVE

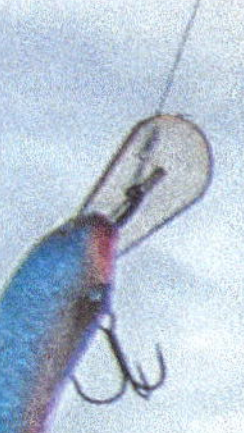

FISHING AND THE FUTURE

Whether you are fishing with a group of friends, with your family, or on your own, fishing is a popular way to enjoy nature, local wildlife, and the beauty of the water. However, all anglers should also respect the environment and follow the local laws and regulations. Ethical, responsible

Always obey local laws and posted warnings at fishing locations.

fishing is necessary to protect the fish species and the habitats they live in.

Declining fish populations and mercury contamination are two major problems that should concern every angler. They are serious problems, but they do have solutions.

OVERFISHING AND POLLUTION

Fish populations are in decline because of overfishing and pollution. Preventing both is within human control, so it is important for every angler to help prevent these problems and be a part of the solution.

One reason fish populations are in decline is due to overfishing. When a species is fished too much, there are not enough fish left to reproduce and keep the population at a steady or increasing level. Overfishing is a particular problem for commercial fishing of popular eating fish, such as tuna, swordfish, and cod. However, even sport fish, such as bass, are in decline. The major

Pollution on land can enter waterways and contaminate the water and the fish. Be sure to clean your fishing area when you are done. Consider cleaning up garbage others have left behind, too.

way to counteract overfishing is to always practice catch-and-release fishing. Take care of fish so that they can return to the water in healthy condition and continue to reproduce.

The other contributing factor to declining population is pollution. Some chemicals that enter the water can kill fish immediately. Other chemicals and pollution reduce the amount of oxygen in the water. Over time, the oxygen level can become so low that fish are not able to survive.

Although the major way large amounts of contaminants enter bodies of water is through pollution from factories or manufacturing plants, individuals can help prevent pollution. Don't dump anything in bodies of water. When fishing, don't leave garbage or

CAREER PROFILE

Some people turn their love of nature and fishing into a career. Federal wildlife officers work for the US government to protect America's wildlife, habitat, and treasured federal lands and waters. They help protect 850 million acres (344 million hectares) of lands and waters. This land contains about 392 threatened and endangered species, 700 species of birds, more than 1,000 fish species, and iconic species such as the American bison, elk, grizzly bear, and more.

A typical federal wildlife officer might teach classes on hunting or fishing safety. They perform wildlife counts to keep track of animal populations. Officers also meet with state officers to plan conservation and wildlife missions. For people who love the natural environment, becoming a wildlife officer can be an excellent career choice.

other pollutants behind that might find their way into the water. Learn about local companies and potential pollutants that might harm the water. Encourage local officials to make stronger anti-pollution laws and regulations.

METALS IN THE FISH

Mercury is a chemical that enters water supplies mainly from mercury particles in the air. The particles are then washed into the water by rain. Once in the water, mercury enters the bodies of fish and aquatic animals. Mercury remains in the organism's body, and if that animal is eaten, it enters the body of the animal that ate the contaminated organism. This occurrence means that fish (or animals) that eat smaller organisms have higher mercury concentrations in their body. When humans eat these fish or animals, mercury contaminates our bodies as well.

Because many types of bass eat lots of smaller organisms, they often contain high concentrations of mercury. Largemouth, smallmouth, and striped bass are on the US Environmental Protection Agency's list of fish that should not be eaten by children and pregnant women. Teens and adults should only eat these types of fish occasionally.

HOW TO HELP

There are many things individuals can do to protect the environment and help keep fish populations healthy. First, fish responsibly. Pay attention to—and carefully follow—all regulations about where and when to fish. Consider always practicing catch-and-release fishing. Don't leave garbage behind in a fishing area, and

Enjoy your next fishing outing by following local fishing rules and sharing your experience with friends and family.

collect the garbage others may have left behind. When boating, be careful not to harm the natural environment. For example, avoid damaging aquatic plants with a boat's motor, and don't spill gas or oil.

Individuals can also educate friends and family about environmental problems that affect fishing. Start a club at school or in the community to promote safe fishing practices. Or consider a career as a fish and wildlife officer. Fishing can be an enjoyable and relaxing pastime. If you fish responsibly, bass fishing can be a rewarding, lifelong hobby.

GLOSSARY

angler A person who fishes.

brackish Partly fresh, partly salt water.

contaminant A substance that adds impurities.

current Water moving continuously in a particular direction.

fiberglass A material made of glass particles fused together.

graphite A material made up of carbon fibers.

hatchling A small fish or animal recently hatched from an egg.

humane Showing sympathy or concern for others.

lactic acid An organic material that is created when carbohydrates are broken down in the body.

mimic To copy or imitate.

monofilament A single strand.

mucus A slick substance produced by the body to moisten and protect.

navigational device An instrument that aids in figuring out position, course, and distance traveled.

nylon A strong artificial substance.

organism A living person, plant, or animal.

predator An animal that kills other fish or animals.

reel A device set on the handle of a fishing pole to wind up or let out the fishing line.

FOR MORE INFORMATION

Association of Fish and Wildlife Agencies
1100 First Street NE, Suite 825
Washington, DC 20002
(202) 838-3474
Website: https://www.fishwildlife.org
Facebook: @FishWildlifeAgencies
Twitter: @fishwildlife
The Association of Fish and Wildlife Agencies represents North America's fish and wildlife agencies and promotes management and conservation of fish and wildlife.

The Bass Federation
5998 North Pleasant View Road
Ponca City, OK 74601
(580) 765-9031
Website: http://bassfederation.com/tbf-youth/tbf-junior-anglers
Facebook: @TheBassFederation
Twitter: @bassfederation
The Junior Anglers program of the Bass Federation allows anglers aged eleven to eighteen to be part of local bass fishing clubs.

Canadian Sportfishing Industry Association (CSIA)
171 Rink Street, Suite 102
Peterborough, ON K9J 2J6
Canada
(705) 745-8433
Website: http://www.csia.ca
The CSIA works to preserve fishing opportunities for Canada's residents and visitors. This is accomplished through education and management.

FishAmerica Foundation
1001 North Fairfax Street, Suite 501
Alexandria, VA 22314
(703) 519-9691
Website: http://fishamerica.org
Facebook: @FishAmerica
Twitter and Instagram: @fafgrants
FishAmerica is working to preserve sport fish populations and keep waters healthy in the United States and Canada.

Fisheries and Oceans Canada
200 Kent Street, 13th Floor
Station 13E228
Ottawa, ON K1A 0E6
Canada
(613) 993-0999
Website: http://www.dfo-mpo.gc.ca/index-eng.htm
Facebook and Instagram: @FisheriesOceansCAN
Twitter: @FishOceansCAN
This organization is Canada's official site for information on all fish-related topics in Canada.

Lady Bass Anglers Association (LBAA)
Website: http://www.ladybassanglers.com
Facebook, Twitter, and Instagram: @LadyBassAnglers
LBAA's goal is to create professional fishing programs for women and to get more young women interested in the sport.

National Oceanic and Atmospheric Administration (NOAA) Fisheries Service
1315 East West Highway
Silver Spring, MD 20910

Website: https://www.noaa.gov/fisheries
Facebook and Twitter: @NOAA
The NOAA Fisheries Service is in charge of protecting the marine resources of the United States.

US Environmental Protection Agency (EPA)
Ariel Rios Building
1200 Pennsylvania Avenue NW
Washington, DC 20460
(202) 272-0167
Website: https://www.epa.gov
Facebook, Twitter, and Instagram: @EPA
The EPA is in charge of protecting humans and our natural environment—air, water, and land.

FOR FURTHER READING

Beard, Daniel Carter. *Do It Yourself Bushcraft: A Book of the Big Outdoors.* Mineola, NY: Dover Publications, 2017.

Berger, Karen. *Essential Knots: Secure your Gear When Camping, Hiking, Fishing, and Playing Outdoors.* Cambridge, MN: Adventure Publications, 2019.

Bourne, Wade. *Basic Fishing: A Beginner's Guide*. New York, NY: Skyhorse Publishing, 2015.

Cermele, Joe, and the editors of Field & Stream. *The Total Fishing Manual: 317 Essential Fishing Skills.* San Francisco, CA: Weldon Owen, 2017.

Gallagher, Tim. *Born to Fish: How and Obsessed Angler Became the World's Greatest Striped Bass Fisherman.* Boston, MA: Houghton Mifflin Harcourt, 2018.

Karas, Nick. *The Complete Book of Striped Bass Fishing: A Thorough Guide to the Baits, Lures, Flies, Tackle, and Techniques for America's Favorite Saltwater Game Fish.* New York: Skyhorse Publishing, 2016.

Llanas, Sheila Griffin. *Largemouth Bass.* Minneapolis, MN: ABDO Publishing, 2014.

Lutz, Chris. *Bass Fishing: A Guide to Mastering Freshwater Bass Fishing Techniques.* CreateSpace Independent Publishing Platform, 2016.

Peluso, Angelo. *An Angler's Guide to Smart Baits: Tips and Tactics on Fishing Twenty-First Century Artificials.* New York, NY: Skyhorse Publishing, 2018.

Root, James. *Smallmouth Bass Fishing for Everyone: How to Catch the Hardest Fighting Fish That Swims.* New York, NY: Skyhorse Publishing, 2017.

BIBLIOGRAPHY

Burnley, Eric. *The Ultimate Guide to Striped Bass Fishing: Where to Find Them, How to Catch Them.* Guilford, CT: The Lyons Press, 2006.

Clouser, Bob. *Fly-Fishing for Smallmouth.* Mechanicsburg, PA: Stackpole Books, 2007.

George, Jean Craighead. *Pocket Guide to the Outdoors.* New York, NY: Dutton Children's Books, 2009.

Greenland, Paul R., and AnnaMarie L. Sheldon. *Career Opportunities in Conservation and the Environment.* New York, NY: Checkmark Books, 2008.

Hill, Stephen. Field & Stream.com. "Unbreakable: 15 World-Record Catches That May Never Be Beaten." June 12, 2016. https://www.fieldandstream.com/photos/gallery/2016/06/unbreakable-15-world-record-catches-that-may-never-be-beaten#page-14.

Labignan, Italo. *Hook, Line, and Sinker: Everything Kids Want to Know About Fishing.* Toronto, ON, Canada: Key Porter Books, 2007.

Newkirk, Ingrid. *The PETA Practical Guide to Animal Rights.* New York, NY: St. Martin's Griffin, 2009.

Popular Mechanics. *How to Tempt a Fish: A Complete Guide to Fishing.* New York, NY: Hearst Books, 2008.

Rose, David. *The Fishing Boat: Use the Secrets of the Pros to Select and Outfit Your Boat.* Guilford, CT: The Lyons Press, 2009.

Schultz, Ken. *Ken Schultz's Essentials of Fishing.* Hoboken, NJ: John Wiley & Sons, 2010.

Underwood, Lamar, ed. *1001 Fishing Tips: The Ultimate Guide to Finding and Catching More and Bigger Fish.* New York, NY: Skyhorse Publishing, 2010.

US Environmental Protection Agency. *Guidance for Implementing the January 2001 Methylmercury Water Quality Criterion.* Washington, DC: US Environmental Protection Agency, 2010.

US Fish and Wildlife Service. Refuge Law Enforcement. December 8, 2018. https://www.fws.gov/refuges/lawenforcement/career-information.php#.

INDEX

ABOUT THE AUTHORS

Jane Katirgis has a bachelor's degree in biology and a master's degree in environmental science. She is a science editor and author of many nonfiction books and science textbooks. She and her husband enjoy fishing, especially in salt water on the East Coast.

Simone Payment has a degree in psychology from Cornell University and a master's degree in elementary education from Wheelock College. She is the author of many books for young adults.

ABOUT THE CONSULTANT

Benjamin Cowan has more than twenty years of both freshwater and saltwater angling experience. In addition to being an avid outdoorsman, Cowan is also a member of many conservation organizations. He currently resides in west Tennessee.

PHOTO CREDITS

Cover Roy Morsch/Corbis/Getty Images; pp. 4–5 (background) smiltena/iStock/Getty Images; p. 5 Chad Ehlers/The Image Bank/Getty Images; p. 7 Rostislav Stefanek/Shutterstock.com; p. 8 (top) RLS Photo/Shutterstock.com; p. 8 (bottom) Franco Banfi/WaterFrame/Getty Images; p. 9 (top) Tom McHugh/Science Source/Getty Images; p. 9 (center) Will Parson/Chesapeake Bay Program/CC BY-NC 2.0; p. 9 (bottom) Eric Engbretson/US Fish and Wildlife Service/ Bugwood.org/CC BY-NC 3.0 US; p. 11 LanaG/Shutterstock.com; pp. 12–13 David Doubilet/ National Geographic Image Collection/Getty Images; pp. 16–17 Bure_NS/Shutterstock.com; p. 19 Jekurantodistaja/iStock/Getty Images; pp. 22–23 ARENA Creative/Shutterstock.com; pp. 24–25 GeorgePeters/E+/Getty Images; pp. 28–29 cpaulfell/Shutterstock.com; pp. 30–31 Hero Images/Getty Images; pp. 34–35 Hailshadow/iStock/Getty Images; pp. 36–37 Courtesy Pennsylvania Fish and Boat Commission; p. 41 Fabien Monteil/Shutterstock.com; pp. 42–43 Connor Howe/Shutterstock.com; pp. 44–45 Matt Jeppson/Shutterstock.com; p. 48 sta/ Shutterstock.com; p. 49 (foreground) Drake Fleege/Alamy Stock Photo; pp. 50–51 j-ezlo/ iStock/Getty Images; pp. 54–55 scottgardner/Pixabay; chapter openers (background) mario31/iStock/Getty Images.

Design: Michael Moy; Layout: Ellina Litmanovich; Photo Researcher: Nicole DiMella